Disclaimer

The information contained in this book is for general informational purposes only. While the author has made every effort to ensure the accuracy and completeness of the information provided, they assume no responsibility for errors or omissions or for any outcomes resulting from the use of this information.

Published by

Grégoire MARCHETAUX, through Amazon.

Edition

First Edition, 2023
ISBN: 979-88-52232-85-4

Dedication

To my wife for her endless support and lively encouragements, as well as for her understanding about my regular overtime.

To my son, who I hope will also be curious and eager to share, but who fills me with joy regardless.

To my parents, who always supported me and enabled me to get a job I love.

Special thanks

To my reviewers, technical or not, who spent time helping me adjust the accuracy-entertainment balance for this book.

TABLE OF CONTENTS

TABLE OF FIGURES

FOREWORD

From the first humans to venture in Space or set foot on the Moon, the ambitious next steps on Mars, or the continuous unravelling of universal mysteries by space telescopes, planetary rovers and interplanetary probes... to the advent of ubiquitous planetary services, space missions have continuously pushed the boundaries of human achievement.

Yet, behind these momentous milestones, captivating discoveries and new mysteries, lies a vast ecosystem of dedicated individuals who work tirelessly to make it all possible. In this book, we uncover the untold stories of those who contribute their expertise and unwavering determination to the space industry – from an insider's perspective.

This book paints an impressionist picture of what lies behind the scenes of the Space industry. Using carefully-picked common questions you might have already asked yourself, it will start with broad strokes and will progressively add more and more details:

- "What does the Space industry enable? What makes you get up in the morning?"
- "What is it meant by the Space industry? Where could I imagine to work?"

9

- "What do you typically work on?"
- "Why is it challenging? How is it different from other industries?"
- "How are companies organised? In which team could I imagine myself working?"
- "What roles typically exist? Who could I imagine to be?"
- "How do missions and projects unfold? How do they influence the day-to-day work?"
- "What would it be like if I worked on a typical project?"

This is my personal attempt at inviting you on a "private tour", starting from high in the clouds for a birds-eye view of the industry, and moving closer and closer to my desk.

Although I am not omniscient and can only extrapolate my own experience, I have worked in Space companies from industry leaders to startups, in UK, France, and Canada, in different types of teams and, and throughout all stages of development, as well as with a lot of the major players, which leads me to believe this book should be sufficiently accurate for its modest intent and purpose of opening up to the public. For the same reasons, I expect it to be more accurate about research, development, and production, compared to other areas, but the entire Space industry conveniently revolves around this workflow.

I liked the idea of writing a one-size-fits-all book: for my wife, my son, my family, my friends, and all others who are curious, but also for those who contemplate a career in the Space industry and would like to decide if it is for them, prepare for an interview, or even jumpstart the first days on the job. I hope you will like the balance I have set to meet this challenging objective.

As the saying goes,

> *"Everyone you will ever meet knows something you don't."*

So let us meet today.

I genuinely hope to learn something from you too one day.

PURPOSE

What does the Space industry enable?

What makes you get up in the morning?

Humans have several environments they can leverage: underground, land, oceans and seas, air, and Space.

Space has the unique advantages of allowing, among others:

- Long-term microgravity
- A much broader view on our entire planet
- A much crisper view of outer Space, beyond the atmosphere
- Physical access to untapped knowledge and discoveries of our Universe

This provides us with a literally infinite potential of applications, only limited by our imagination. In only 62 years or about 3 generations, the following applications have not only been imagined but also brought successfully into existence, and are still evolving.

- Space exploration
- Space utilisation
- Space science & technology[1]

More specifically:

- Space-based research
  - Microgravity for medicine, material sciences, biology…
  - Experimental physics

© ESA, 2010

Figure 1: International Space Station (1st launch in 1998)

- Space exploration

---

[1] Technology is created from applied science and represents tools that engineers can use

- Manned missions
- Planetary probes and rovers[2]
- Interplanetary or even interstellar probes
- Associated R&D such as terraforming, resources extraction, local manufacturing, faster propulsion[3]...

© ESA, 2013 (both images)

Figure 2: Render of concept of 3D-printed habitats on the Moon and demonstrated 1500kg building block

---

[2] Semi-autonomous robots

[3] Some science-fiction-like concepts are under study but, to this day, only work on paper - and would take decades or even centuries to materialize (e.g. fusion drives...), if at all for the most groundbreaking (e.g. blakhole drives, warp drives...)

- Space observation
    - Astronomy and cosmology
    - Asteroids threat early warning
- Defence and security
    - Intelligence, surveillance, and reconnaissance
    - Satellite proximity protection
    - Threat interception

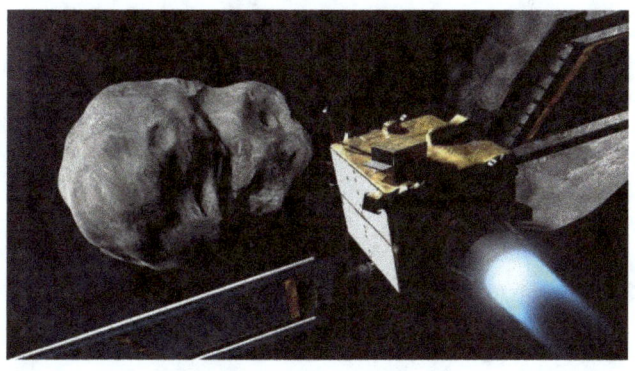

© NASA, 2022

Figure 3: DART (Double Asteroid Redirection Test) interceptor mission (flown in 2022)

- Planetary services
    - Global navigation

- o Assets tracking
- o Search & rescue
- o Environmental monitoring
- o Weather forecasting
- o Smart agriculture
- o Land use and urban planning
- o Natural resource management
- o Disaster monitoring and early warning
- o Telecommunications (TV, radio, telephone, internet, …)
- Space tourism
- Research and development in alternative launchers (e.g. slingshots, railguns, orbital lifts) and rocket technology (e.g. reuseable rockets, or greener fuels)
- In-orbit services (autonomous) are under study. In increasing order of complexity:
  - o Re-fueling and deorbiting
  - o Some repair and maintenance
  - o In-orbit assembly of mega-structures (e.g. SpiderFab[4])

---

[4] I know firsthand that at least two other companies develop their own entirely different technology for this application

© ESA, 2016

Figure 4: Render of a concept of a deorbiting servicing satellite

The very fact that we could just as well be imagining or implementing the next application, or developing those that are already shaping all our lives, is purpose enough in my book (pun intended).

Although we do not chose what mission comes on our desk, they always keep coming in[5] and there are more hopeful, promising, as well as fascinating missions than there is time for us at work.

I am aware the Space industry has a significant impact on the environment, but I personally believe

---

[5] Thanks to everyone in the chain, c.f. "Ecosystem"

that the net effect can already be positive, and that it is all a matter of what applications are developed. "Already", because there is also a high potential for improvement in launches into Space -which until recently essentially stayed the same for decades-, and that potential is starting to be tapped into. As R&D breakthroughs accelerate and get implemented, the net effect will get even more positive.

Everyone in the Space industry has *some* power to shape the collective approach, strategy, and ultimate impact of leveraging the incredible potential of Space, to make the World a better place to live in.

That is what gets me up in the morning, at least.

That, and the incredible everyday-learning experience.

ECOSYSTEM

What is it meant by the Space industry?

Where could I imagine to work?

The Space industry is a small world in that it is common to come across the same company names, and even faces in a career. However, the overall reach of the Space sector is large, and the landscape of actors within it, varied. In parallel with the natural world, the Space sector thrives on the symbiotic relationship between its contributors, each of whom brings something special to the table.

Landscape of actors

The current landscape can be mostly described in the following way:

- Academia, where research often drives missions
  - Dedicated space research centers and laboratories (e.g. European Space TEchnology Center)

- o Universities and other general research locations, potentially with specialised departments (e.g. Caltech)
- o Space education, training, and outreach (e.g. US Air Force Test Pilot School, Women In Aerospace association)
- Governmental entities, which control and regulate missions, and the sharing of Space
  - o International groups of national governmental entities (e.g. European Space Agency)
  - o National agencies, at the helm of the Space strategy of a country (e.g. Canadian Space Agency)
  - o National military & defence, who may be utilising Space but also regulate it much in the same way they do for sea and airspace (e.g. « Commandement de l'Espace »)
  - o Regulatory bodies who ensure regulations that implement Space share treaties are followed (e.g. for spots in orbit with the International Telecommunications Union, or for radio spectrum with the Federal

Communications Commission). Note that regulatory bodies may be private, and only mandated by the governments.

- Non-space commercial customers, who pay for turnkey missions to generate revenue
  - Telecommunications operators, who basically launch networks of pseudo-communication towers in Space (e.g. Intelsat)
  - Other Space applications service providers, who sell ways to use spacecraft data (e.g. Planet labs)
- Commercial space companies, i.e. industrials from the Space sector
  - Spacecraft manufacturers (satellite, launcher, probes and robot manufacturers e.g. Airbus or MDA, space tourism companies e.g. Virgin Galactic)
  - Payload manufacturers (e.g. RAL Space[6])
  - Ground system manufacturers (e.g. L3Harris)

---

[6] RAL might not be as well-known, but they did take part in many major missions. Disclaimer: I worked there

- o Operators, who buy or rent spacecrafts and/or infrastructure and offer them as a service (launch e.g. Arianespace, and ground stations e.g. Telespacio Vega)
- Test houses, where tests are conducted if subcontracted (e.g. NTS)
- General industry, because some components are ordered from leaders of the general industry, tested and potentially adapted for Space (e.g. Maxon motors in Curiosity)

Jump to §"Products and services" if some of the products or services appear vague.

Some of these actors may actually be part of a single company or corporate group. Keep in mind also that the space industry is a complex and ever-evolving landscape, with new actors emerging and existing ones evolving[7].

Strategic facilities

Some of these actors have strategic facilities without which projects could not be completed, such as the ones listed below. We will talk about why they are

---

[7] Especially now, with the rise of SpaceX that proved long-established industrials that the "NewSpace" approach could rival traditional approaches.

needed shortly, in §"Challenges & characteristics".

- Assembly and integration cleanrooms: where the Space systems are assembled, integrated into a larger system, and tested prior to launch.
  - There, the particulate contamination for:
    - Sensitive equipment such as optics, detectors, and small orifices
    - Planetary protection (no contamination of planets explored)

    is controlled using:
    - Non-turbulent flow of air from the top
    - Air Quality / particle count sensors
    - Attire

    There are a few classes of cleanliness, which can require simple gown, headnet, beard cover and overshoes (usually sufficient for non-optical satellite equipment), or a full-body suit.

- ElectroStatic Discharges[8], harmful and often lethal to electronics, are prevented using:
  - Dissipative[9] floor
  - Dissipative workstation mats, chairs
  - Wrist straps and shoe straps connecting the skin to the above
  - Strap testers
  - Air ionizers in certain cases

©NASA, 2022

Figure 5: 6U Cubesats under functional testing

---

[8] Similar to zaps you get coming out of your car
[9] "Electrically leaky to ground"

- ○ Changing rooms, and equipment cleaning airlocks, ensure a safe transition

© ESA, 2002

Figure 6: ENVISAT under integration at ESTEC (flown in 2002)

- Testing facilities where the application's environment is able to be simulated:
  - ○ ElectroMagnetic Compatibility and Radio Frequency test facilities or houses: these facilities have walls

which are designed to absorb all radio waves, in or out, to simulate ("quiet") Space, and are equipped to test emissions and susceptibility of radio equipment to requirements (to a standard, or for performance assessment).

Figure 7: Anechoic chamber for RF testing at ESA's ESTEC

○ Thermal Vacuum chambers: these chambers are able to be pumped down to a suitable vacuum, while heating or cooling the mounting location of the equipment to test, and even sometimes the inner walls acting as a radiation environment. Specific profiles are followed.

TVACs are slow, and impractical for anything that has to stick out from the chamber (e.g. wires), so it is common to resort to thermal chambers (basically a combined oven and fridge) in early stages of development.

The temperature of the equipment to test is monitored throughout, in many key locations.

© ESA/Lightcurve Films, 2021

Figure 8: JUICE in the Large Space Simulator, Europe's largest TVAC, at ESA's ESTEC

o Vibration & shock facilities: the equipment to test is mounted on or to, basically, a giant speaker. Said *shaker* is controlled to specific profiles. Accelerometers are used in key locations to monitor critical resonances. For shock testing of fragile parts, the equipment is hit by a hammer-like part of which the repeatable blow is first experimentally tuned to the desired profile on a dummy part.

Figure 9: Electronics under vibration test, using a slip table

- o Radiation test facilities: components (usually electronics) are bombarded by a radioactive source calibrated to output a stable flow of high-fixed-energy particles of specific types, defined by the requirements.
- o Simulated field trials: for some spacecrafts, the above is not describing the full picture of the application's environment. For planetary robotics, for example, field trials are simulated using a specially-designed rocky landscape with controlled lighting to test navigation and locomotion.

© Airbus Defence and Space Stevenage, 2014

Figure 10: Martian rover field trials facility

- **Spaceports**: where the systems are actually launched (and now, can be recovered). Note that launch facilities include Assembly Integration and Tests facilities for the launcher itself, and for integration of the payload(s) in the launcher.

© NASA, 1969

Figure 11: Saturn V is rolled out of the Vehicle Assembly Building and down the crawlerway to the launchpad

- **Astronauts training centers**: where astronauts are trained, evaluated for their ability to survive and carry out missions, and selected

Figure 12: Underwater astronauts training at NASA's Neutral Buoyancy Laboratory

- **Ground segment**: the complementary part to the flight segment (the spacecraft(s)), in operation, comprised of:
    - Ground stations: where the signals to/from the flight segment are sent or received and dispatched for processing. Ground stations exist across the planet for maximum coverage, and are generally linked in networks to connect control centers and data processing centers with the flight segment.

- o Control centers: where the flight segment is operated and monitored. They may be located in a ground station, or in a launch facility to monitor a launcher's flight (which is always autonomous). Payload data may be forwarded to data processing centers.

Figure 13: One of NASA's Flight Control Rooms in Houston

Interactions

The space sector forms a vibrant tapestry of interactions, in which the collective efforts of all its actors drive its growth and progress. Contracts and agreements govern interactions, but also reflect mutually advantageous relationships. The following diagrams show a glimpse of a web that would otherwise be filled with arrows.

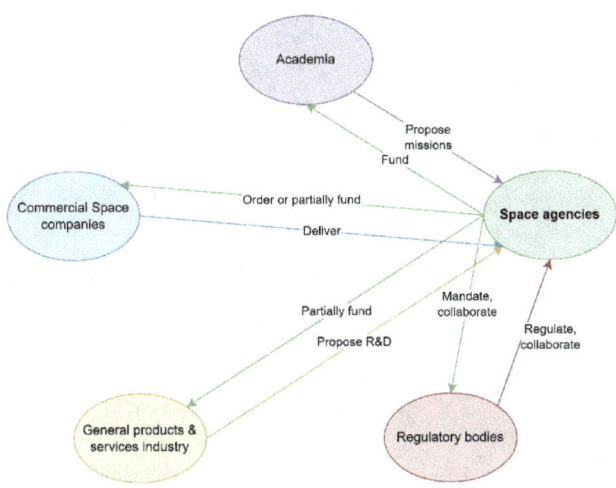

Figure 14: Example of interactions as seen from Space agencies

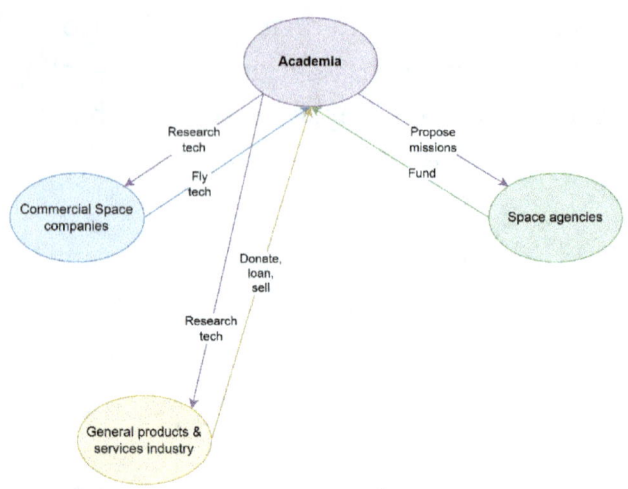

Figure 15: Example of interactions as seen from Academia

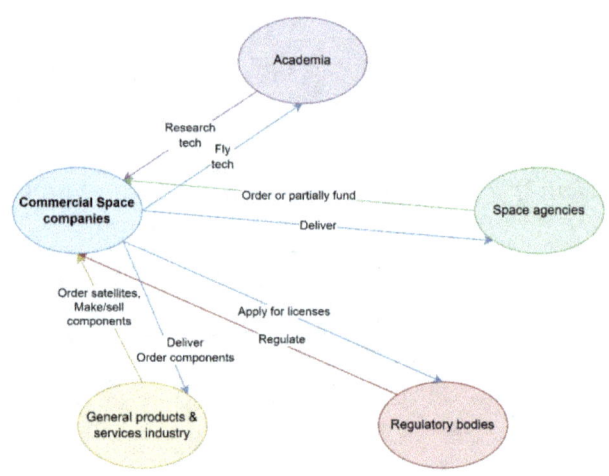

Figure 16: Example of interactions as seen from commercial Space companies

34

As you can see, commercial Space companies could take orders from the general public services industry, or from academia via Space agencies which are funded by countries. Industrials who may be emerging can even be offered to demonstrate their technologies in Space, by having their demonstrator hosted on a government-funded technology demonstration mission or on a commercial satellite with some room to spare (for a price).

Some say that the missions funded by space agencies are a waste of tax revenue but firstly, the budget of national space agencies is actually negligible compared to other governmental programs[10]. Secondly, high efficiency of the return on funding seems to have been demonstrated, based on the achievements mentioned. Thirdly, many jobs are created; in particular, production is automated very little due to the low volumes. Finally, the economic benefits shine not only on the entire sector but also far beyond, through ricochet effect.

---

[10] E.g. between 0.3 to 0.5% of the total US budget gets allocated to NASA, c.f. https://www.planetary.org/space-policy/nasa-budget

PRODUCTS & SERVICES

What do you typically work on?

As applications always keep evolving, products made and services offered by the Space industrials evolve with them – nonetheless, the following list should cover most of them for the foreseeable future.

Spacecraft

Typical Space systems include:

- **Satellites** are like cannonballs shot into the air which never stop falling, and which can perform useful tasks from their resulting orbit.
  - They can be orbiting:
    - Earth (e.g. Hubble)
    - Our Moon (e.g. LRO),
    - Mars (e.g. MRO[11])

---

[11] Mars Reconnaissance Orbiter was extremely useful to the guidance of rovers on Mars: rovers match what they see with what MRO sees from orbit to locate themselves – c.f. Visual Odometry.

Figure 17: Mars Reconnaissance Orbiter (left, launched in 2005) useful to many Mars missions including Ingenuity (right, launched in 2020)

- A comet (e.g. Rosetta)
- Any other planet (e.g. JUNO around Jupiter, Cassini-Huygens around Saturn)

© ESA, 2001

Figure 18: ROSETTA, launched in 2004, dropped the lander Philae on Comet Churyumov–Gerasimenko in 2014

- The Sun (e.g. STEREO, SOHO)
- Lagrange points[12] (e.g. JWST, GAIA)
  - They follow an orbit in a wide range of altitudes (from 100km to around 900 000km of "influence" for the Earth) and other orbital parameters[13] such as in (for Earth):

---

[12] Specific points in Space where the attraction from celestial bodies is minimal, because they cancel each other there

[13] Describing any ellipse of which one of the two "centers" of the ellipse is something in this list

- Low Earth orbits: until around 1600km of altitude (e.g. for inexpensive launch costs). Short life due to drag.
- Sun Synchronous Orbits: nearly-polar orbit in the 600-800km altitude range designed to always pass over at the same local time (e.g. for observation)
- Geostationary orbit: equatorial orbit at 36000km of altitude, where the spacecraft rotates at the same rate as the Earth, and is stationary with respect to ground. (e.g. for large telecommunications satellites)
- Highly elliptical orbits (e.g. Halley's comet) spend more time above certain regions than others, albeit further away.
- Geostationary Transfer Orbit: specific case of elliptical orbit into which launchers launch satellites from low altitudes, to eventually reach geostationary altitude. Satellites "maneuver"

on the other side ("apogee") to lock into orbit.

- o They can be:
  - Single satellites (e.g Envisat)
  - In constellations or formations (e.g. GPS, EDRS, Swarm)

© ESA, 2013

Figure 19: Artist view of the deployment of the formation-flying SWARM satellites (launched in 2013)

- o They can vary wildly in size:
  - From Cubesats as little as a cubic tissue box deploying card-sized subsatellites (e.g. KickSat-1)

40

- To space stations made of individual modules, launched at different times and assembled in orbit (e.g. MIR, ISS, Tiangong)[14]

- **Probes** are unmanned systems sent on a specially-designed fixed trajectory that can terminate around a planet (where they would brake and turn into a satellite), or without ever stopping (e.g. Voyager-1 and Voyager-2).

© ESA, 2010

Figure 20: Voyager-2, launched in 1977 and now well beyond our Solar system

---

[14] There even is a dummy-driven Tesla car orbiting around the Sun...

- **Landers** are elements of satellites which are able to land on the body around which they initially orbited (e.g. Lunar Module or Curiosity's Sky Crane[15])

Figure 21: Apollo Lunar Module (left, launched in 1969) and Curiosity and its sky crane (right, launched in 2011)

- **Exploration robots** are semi-autonomous robots which are able to autonomously follow carefully-prepared plans sent from Earth because remote control would be impractical (due to the round trip delay of the communications that "only" travel at approx.

---

[15] Not a lander per se, but lands the rover nonetheless

the speed of light) (e.g. rover Perseverance, helicopter Ingenuity)

- **Space vehicles** are modular transport vehicles that ferry cargo or crews, between a launcher's drop-off and most often a space station in Earth's orbit (e.g. the Space shuttles, or the capsules Soyuz, ATV, Dragon, Orion), but in the past also to the Moon, and in the future to Mars or beyond.

© NASA, 1999; © SpaceX, 2021

Figure 22: Space Shuttle servicing Hubble (left, 1999) and SpaceX Crew-1 Dragon capsule docking to the ISS (2020)

- **Launchers** are transport vehicles that send all systems mentioned above into Space, either using multistage rockets (e.g. Ariane-5, Falcon-9, Saturn-V) or potentially alternate concepts under study (e.g. spaceplanes, railguns, slingshots, space elevators…). They can send one or multiple payloads, in one or multiple orbits, through the use of dispensers and additional stages (i.e. "mini-launchers") - in the same fashion as a bus.

© NASA, 2018; © NASA, 2023

Figure 23: Saturn-V, still the largest rocket ever built (1st launch 1967) will be dethroned by the stacked Starship from SpaceX

All those systems are comprised of:

- The platform, of which the only purpose is to allow the payload to operate. Imagine the platform like a car with a driver who can never stop. Manufacturers usually have a platform products range they reuse and adapt (e.g. Airbus's Eurostar3000)

- The payload, which actually does what the mission is for. In the previous analogy, the payload is someone with a mission, driven around by the driver. Like a photograph. Payloads are usually custom-made to each program or mission because of the uniqueness of the mission or the challenge in improving the performance of already cutting-edge technology. The exception is the launcher, for which the payload is the spacecraft(s).

Spacecrafts' **payloads** are simply executing a mission-specific task -any task-, but they are often instruments. **Instruments** are extremely advanced sensors of which the working principles were imagined and developed by researchers (not necessarily in the Space industry), and then adapted for Space in collaboration with industrials from the sector.

They vary wildly in nature:

- High-resolution imagers, in or out of the visible light wavelengths, including sets of imagers which combine into a more powerful imager by the use of a process called interferometry (e.g. JWST)

© NASA, 2019

Figure 24: The James Webb Space Telescope under testing of deployment (launched in 2021)

- Coronographs[16], e.g. to predict bursts of particles from the Sun, which are harmful to satellites (e.g. in Solar Orbiter)
- Spectrometers[17], to analyse the composition of matter (e.g. in Curiosity)

---

[16] Mask a star to look at the variations of its crown
[17] Shine light through matter and analyse what comes through

- Radars, laser altimeters, and other echo-based sensors, e.g. for all-weather imagery, oceanography… (e.g. Sentinel-3)

© ESA, 2021

Figure 25: Solar Orbiter, between Mercury and the Sun (launched in 2020)

- Radiometers, magnetometers and other environmental sensors, e.g. to probe the magnetosphere for geomagnetic storms (e.g. Cluster)
- Radio receivers, e.g. to listen for radio signals (e.g. Voyager)
- Gravitational wave detectors (LIGO) and other experimental physics proofs of concept

Other payloads include:

- High-power transmitters with high-gain receivers and antennas (e.g. Jupiter-3)
- Sampling equipment such as drills, robotic arms, and collection appendices (e.g. Perseverance & MSR, or DEXTRE, Canadarm)

© NASA, 2021

Figure 26: Mars Sample Return lander launching the Mars Ascent Vehicle (cancelled), as seen from Perseverance (launched in 2021)

- In-situ experiments (e.g. microgravity research on the International Space Station, or MOXIE which demonstrates Oxygen generation on Mars)

Figure 27: ISS's Special Purpose Dexterous Manipulator (DEXTRE) (launched in 2008)

Platforms and payloads alike are complex systems, which are typically divided in subsystems.

Subsystems typically are:

- Structure: the glue which holds everything together even through the unforgiving launch environment (c.f. §"Challenges and characteristics")
- Thermal control: regulates the temperature of key components within their operating range

- Power: provides a steady supply of power to all electrical components (from the Sun, or from the heat of radioactive material)
- Communication: receives commands from the Earth, and transmits back down data from the platform and the payload
- Harness: designates all wires and cables connecting all electrical subsystems together
- Attitude and Orbit: determines and controls the trajectory and the orientation of a satellite or probe, to follow its intended flight plan
- Guidance and Navigation: is basically for rovers what you instinctively do when hiking on unstable rocky terrain: know where you are, and how to get where you want to go without falling.
- Data handling: the brain, the intelligence, the memory.
- Mechanisms: anything that moves. E.g. to separate rocket stages, hold and deploy appendages post-launch (where space was limited), augment the field of view of instruments by pointing or scanning, change filters, act on the immediate environment (for drilling, sampling etc.)...

- Propulsion: thrusters and engines, for attitude and orbit/flight path control.

© Airbus Defence and Space, 2017

Figure 28: One of the two articulated ion thrusters for Electric Orbit Raising of Airbus's telecoms platforms (since 2017)

- Crew support equipment: necessary to keeping astronauts alive – such as EVA[18] spacesuits, life support, habitat etc.

---

[18] Extra Vehicular Activities

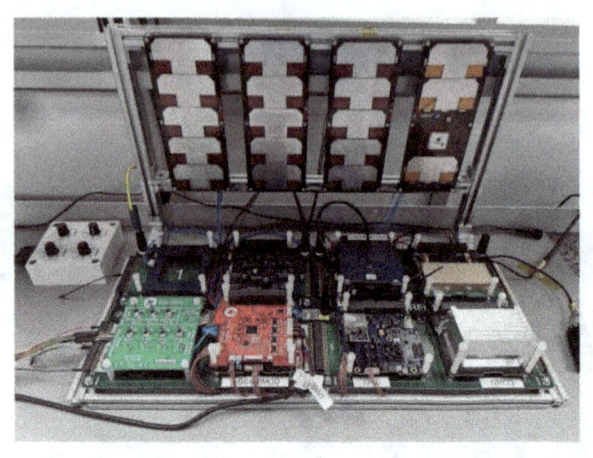

Figure 29: Cubesat "Flat Sat" where all its subsystems are laid flat (EIRSAT-1, launch scheduled in 2023)

All the aforementioned spacecrafts are considered part of the **flight segment**. They actually have numerous support systems.

Ground systems

Be sure to have read §"Ecosystem Facilities" for a description of what ground system facilities are.

Ground systems include **Ground Support Equipment**, which designate equipment a Space system needs on the ground at any stage from development to launch. They can be of any nature (e.g. electrical, mechanical, fluidic, optical) and can for example facilitate:

- Handling (including during assembly and integration) and transport (e.g. for satellites weighing as much as a truck)
- Actual assembly or integration (e.g. for precise alignment)
- Functional or performance tests
- Calibration (e.g. for instruments)
- Launch preparation (e.g. filling)

Finally, both the flight and ground segments will be constituted of components (hardware and software), which in turn will be made of individual parts – all of which are made by a variety of manufacturers including from the global industry. Many people eventually contribute to a Space system.

© ESA, 2016

Figure 30: Electrical Ground Support Equipment for pre-launch checkout of the rover ExoMars 2016

© Lockheed Martin, 2021

Figure 31: Mechanical Ground Support Equipment enables tests of LUCY's solar array deployment (launched in 2021)

Services

Not everyone in the Space industry as defined in §"Ecosystem" works on spacecraft and on ground systems: an acceptable definition in this context is that services cover all work which does not result in the production of spacecraft and ground systems, which means service providers may operate them or turn their outputs into one of the applications mentioned in §"Purpose".

Aside from potentially subcontracted Operations, service providers generally are either the customer which has ordered the mission or program (e.g. telecommunication operators which amount to a large chunk of business for satellite manufacturers), or an unrelated company buying the data to be able to offer a service of interpretation (e.g. image processing companies). Generally, this involves signal, data or image processing, modelling, artificial intelligence etc. which are beyond the scope of this book.

Refer to §"Purpose" for a list of examples of services, since many applications are facilitated by services.

CHALLENGES & CHARACTERISTICS

Why is it challenging?

How is it different from other industries?

It is well-known that consumer products are particularly difficult to design and produce because of the excessively low unit cost targets required to appeal to buyers.

Saving is usually not something that people associate with the Space industry – and to some extent, they are right. However, regardless of how high the budget is, Space remains harsh:

- Missions themselves are challenging to design and follow: contrary to what is shown in science fiction, spacecrafts cannot fly wherever we want. They're actually closer to gliders in that as soon as we launch them, they become like a marble in a marble machine – helpless –, if it was not for the very limited amount of fuel we give them only to compensate for long-term trajectory drifts, and to perform rare manoeuvres in order to change course somewhat. This means absolutely everything is in motion from day one, forever: communication beams, instrument views, incident solar power, radiated heat transfers, potential collisions, the very fabric of gravitational pull, Space weather... Even the push of light itself and the shape of the Earth's crust have a noticeable impact. The rare solutions to these numerous time-varying constraints often still require challenging system requirements (e.g. more fuel, bigger solar panels, bigger antennae, bigger batteries...)

- Spacecrafts have nothing to push against, which means that:
  - They can only move (or maintain speed) by shooting matter the other way[19]. The hotter, denser, and faster – the better, but in any case, at the speeds they already go, much fuel is needed to make significant changes. This means large and heavy tanks.
  - Orientation is very sensitive, just like if the spacecraft was balancing on a

---

needle with its tip inside the craft: any imbalance in mechanisms will try to turn the spacecraft. Thankfully, the high inertia of the spacecraft helps for fast disturbances.

- o Orientation mechanisms (when not fuel-based) are more complex than what could be done on Earth. Bigger and heavier. Moreover, for agility, they often involve spinning devices, whose added (micro)vibrations may disturb instruments.

- Extreme vacuum: the absence of external pressure means:

- o Vacuum allows plasma[20] to exist, which tends to charge conductive surfaces to levels that could result in harmful discharges onto electronics

- o Pockets of gas from the ground which did not fully leak out during ascent do so and expand in Space, potentially condensing on cold optical surfaces, degrading them. Trapping them may also be an issue, since the pressure

---

[20] Another state of matter, basically a non-liquid soup of ions

difference with the surrounding vacuum adds stress on the container.

- o All materials partially evaporate ("outgas"), to varying degrees. Aside from the material loss, particularly problematic for lubrication, the resulting gasses result in the same issues as above.
- o Similar metals tend to cold weld, effectively jamming mechanisms
- o Oddities enabled by microgravity such as "tin whiskers" which grow from unleaded solder, causing shorts
- o Temperatures are less homogeneous because heat convection through air is non-existent (see below)
- Extreme thermal environment: parts are exposed either to extreme heating (e.g. when shooting through atmosphere due to the high friction against the hull at great speeds) or to highly-contrasted sources of radiated heat:
  - o Deep space, at -269°C
  - o The Sun and its high heat flux
  - o Its reflection on the Earth (mostly in low orbits) or own surfaces

All of which also vary with time, just like internal dissipation from the electronics. Things get easily too hot, or too cold, whereas absolutely everything depends on temperature, more or less:

- o Electronics could burn out or at least become de-calibrated and get lower reliability
- o Propellant and other materials could boil and evaporate or freeze
- o Materials contract, expand, and also skew and warp due to mismatched expansion rates in assemblies (e.g. mechanisms jam, alignments degrade). This can lead to immediate issues, or long-term degradations through fatigue by thermal cycling.

- Power self-sufficiency: spacecrafts do not have access to electricity services. They have to include their own powerplant.
- Micrometeorites and space debris: there are countless parts in orbit left from previous launches and missions, and specks of dust, that often travel at tens of thousand kilometres per hour. The more their trajectory is different from the considered

spacecraft at the time of impact, the more difference in speed this makes, and the more dangerous they are. Large surfaces are likely to be degraded by micrometeorites in their lifetime. By a sheer stroke of bad luck, micrometeorites may easily puncture critical components or even astronauts.

- Severe launch vibrations and shocks: launchers expel a phenomenal flow of hot turbulent gases during a launch, which result in high vibrations throughout the structure. The payload(s) shake hard, and across a large range of frequencies. This may excite many resonances[21], potentially damaging things or setting them out of alignment. Brutal mechanisms, when used, also send shockwaves which may be detrimental to fragile components such as optics.

- Intense radiation: electronics, coatings (and therefore thermal performance), and optics are especially sensitive to radiation. Other materials get more brittle as they absorb. Unfortunately, spacecrafts get bombarded from:

---

[21] Think about pushing on a swing, first slowly then faster and faster. There is a frequency where you could easily shoot people off. This is resonance. All assemblies have them.

- o The Sun (especially during CMEs[22])
- o The Van Allen belts around Earth where high energy particles accumulated, trapped by the Earth's magnetic field[23]
- o Galactic cosmic rays coming from deep space
- o Potential nuclear events, if the application is for the military

All the mentioned components degrade as they absorb radiation, in extremely complex mechanisms which are only approximated during tests in laboratories. Electronics, although critical to the mission, is particularly sensitive[24] and can even malfunction before the limit dose when struck by higher energy particles.

- In Low Earth Orbits, oxygen atoms in the upper atmosphere resulting from UV light breaking O_2 bonds, degrade polymers, mirrors, radiators and other surfaces

---

[22] Coronal Mass Ejections. Like bubbles or whiplash in lava which propel energetic particles across Space
[23] Source of the Northern Lights, incidentally
[24] This is mainly why Space has to be lagging behind consumer electronics in terms of computing power – denser is more sensitive

- Limited mass: launchers are still very inefficient ferries. They can only carry a limited mass worth of payload (a fraction of their own mass), and that mass even decreases for more demanding orbits (essentially depending on altitude and inclination[25]). A lighter satellite could also enable boarding other satellites for a multi-launch, which would reduce launch costs (they are very expensive).

- Limited space – resulting in limited power. Launcher fairings have limited space – spacecrafts generally require more space in orbit than what is available in the launcher (e.g. for large solar panels, or antennas), and therefore resort to deployment mechanisms. Power available is always limited regardless, this resource must therefore be carefully allocated and budgeted on board.

- Highly precise instruments and alignments: in the worst-case conditions of all aforementioned obstacles, instruments are generally required to achieve a feat of performance, sensing faint phenomena from massive distances. A not uncommon Earth

---

[25] Angle of the orbit with respect to a standard reference

observation requirement is equivalent to being able to see a hair from 18km away, anywhere in our field of view.

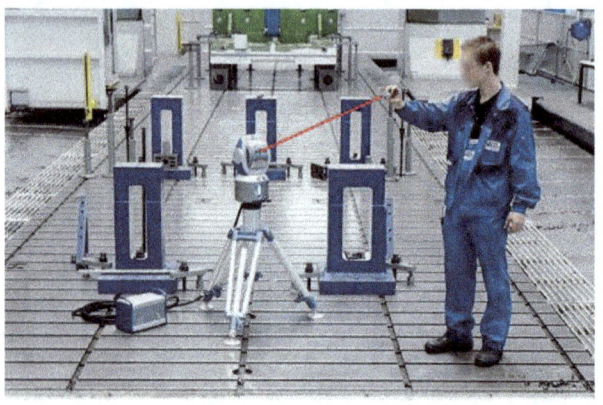

© FARO, via Wikipedia, 2023. Beam added.

Figure 32: Laser tracker and reflective target, used for precise alignments and calibrations

- Communication gaps, attenuation, and delays: communication takes time to propagate through Space, and loses energy as it travels, by collision or dispersion. As a result, transmitters must be powerful, and receivers must be sensitive, but regardless, spacecrafts will likely lose communications in certain conditions (e.g. behind horizon, or in front of the Sun) and experience lags (e.g. a few seconds to the Moon, up to 20min to Mars). This affects manned and unmanned missions alike.

- New worlds mean new environmental challenges. Very thin atmospheres make it harder to brake spacecrafts into orbit, but dense atmospheres are a thermal hazard during braking or landing; low gravity (also on asteroids) means trying to land on a trampoline, but crushing gravity means more stress/load on structures and mechanisms.

- Long life: mission lifetimes vary from a few months to decades, but many Space systems are designed to operate for more than 15 years in the harsh environment described, in order to maximise the return on investment. Voyager-1 is still operational after 45 years.

- No maintenance possible: worst of it all, risks must be extremely carefully accounted for and weighed in, because repairs and maintenance are essentially impossible to this day[26]. Redundant and "just-in-case" measures are added but they result in added

---

[26] The Hubble telescope is a historic example of rare in-orbit repairs, using one of the Space shuttles which are now retired

complexity. Lengthy reliability analyses determine the best compromise.

Specifically for manned missions:

- Life support, pressurised compartments and suits need to be figuratively bullet-proof.
    - Obviously, astronauts cannot breathe in Space nor on other planets so far
    - Water anywhere exposed to vacuum would evaporate (e.g. eyes)
    - Air in blood vessels would boil out, and lungs would dangerously expand[27]
- Radiation poisons astronauts, faster with high solar activity
- Temperature is, as mentioned, difficult to regulate. Especially in atmosphere re-entry vehicles
- Launch and re-entry are hard to tolerate. Astronauts are screened, selected and trained, but there is a limit to what they can be expected to endure and it is not an easy one to meet.
- Astronauts are also selected for their mental strength. Months if not years (or even an

---

[27] So do soft pressurized suits, actually. Neither soft nor hard suits are easy to maneuver.

eternity) far from home and family, trapped all day with the same people in rather precarious conditions, will break most people.

Testing itself is also challenging, because spacecrafts generally cannot deploy on Earth without help (otherwise, they would be overengineered, possibly even impossible to make), and this help introduces errors in the simulation of microgravity.

There are also non-technical challenges, in particular:

- Cost: It is not always about unit cost: Space projects are still expensive. *Very* expensive. The development of a single subsystem or small system typically needs millions of dollars, and a typical program hundreds of millions to a few billions. Why? To avoid spending somewhat less money but still a significant amount, only to eventually launch an overengineered paperweight. Labour cost usually dominates, not only for the development but also for production, which is generally mostly manual labour because of too few units to make[28].

---

[28] A few exceptions, perhaps: Oneweb, Starlink

- o This budget is difficult to secure for the projects
- o Competition results in aggressive "sold" budgets which are then very tight to follow.
- Planning:
  - o Just like costs, competition drives equally tight development and production times: in many applications, delays mean less revenue in operation
  - o Even applications where operation does not generate any revenue, it may be necessary to launch on specific dates (called launch windows) to be able to follow complicated trajectories that use planets to "Spider Man"™ around (gravitational assists) and meet a potential target (rendezvous).
- Politics: sometimes, the way organisations are set up limit options in terms of who gets awarded work (e.g. geo-return of ESA). Some applications (e.g. military) may also require "sovereign" launches, i.e. without dependence on other countries; etc.

All these constraints add up, and on top of already

demanding requirements from the customer, the degrees of freedom are very limited to the point that finding a solution that meets the requirements rapidly eats away the allocated time and money. It is actually very common that compromises across the board are not enough and that requirements negotiations propagate back and forth through the hierarchy until it becomes actually possible to meet them all.

DEPARTMENTS, TEAMS, SPECIALITIES

How are companies organised?

In which team could you imagine yourself working?

Each one of the entities that make up the Space industry is made of many departments, each of which involves many teams, which in turn may define many specialities.

Departments

Every company is free to choose the organisation that best suits its business. Nonetheless, companies tend to have at least, by following mandatory or optional standards, departments related to:

- Human Resources: manages personnel, ensures their rights are respected at all times, and fosters a positive work environment
- Health and Safety: ensures that the company never does anything that could put its employees' health or safety at risk. Trip, fall,

and crush hazards as well as cancerous practices are typically H&S matters.

- Information Technology: manages the electronics and computer systems in the infrastructure of the company, which support its operation. There generally is a team specialised in the security of the systems in place
- Legal: handles everything related to the law, such as contracts, regulations compliance, intellectual property
- Sales, business development, and Communication/marketing: drives business growth by engaging with customers, promoting products and services, and handling the company's image
- Accounting and Finances: manages everything related to the company's finances: past, present, and future.

In addition, Space industrials generally implement:

- Science and technology: researches and develops new ideas and concepts that could be assets in the company's future plans, until they are mature enough for Engineering to take over and use them in new projects.

- Engineering: designs products which meet specific performance, cost, and lead time targets.
    o Design office: self-explanatory
    o Manufacturing, Assembly, Integration[29] and Tests: where the manufacturing, the assembly, and the tests of the products designed in the design office, are done. Due to low volume, building Space products relies heavily on qualified labour. Though usually dominated by operators and technicians, the MAIT department has a particularly high fraction of design engineers within its ranks in the Space industry, possibly because of the extensive planning required to build products which are so expensive.
- Quality Assurance ensures that products and services meet the expected quality, through systematic processes and continuous improvement. It is basically the memory of everything that went wrong at some point (potentially in other companies), and the enforcer of all corrective measures that

---

[29]Assembly into a larger system, usually the satellite

resulted. You would typically hear them asking: "Any anomalies? How/why? What can be improved? Are you following [some standard]?" etc. If you see Quality Assurance and Product Assurance, Product Assurance is closer to product development compared to QA.

- Procurement or Supply Chain: purchases, or centralises purchases for, products from other companies on behalf of other departments in the company. "Make, or buy?" is a question that applies to every component in the products – this department handles the "buy" aspect whereas Engineering does "make".

Operations is another common department, but its meaning varies wildly depending on the industrial entity in which it is implemented. At its core, it could be:

- Operation of the company, by a delegate from the Chief Executive Officer
- Operation of specific infrastructures, such as a launch site

- Operation of spacecrafts[30]

Teams

Every department is generally subdivided into teams, so this section will take the example of the **Engineering** department.

Teams (or groups) are generally consistent with the products and challenges we have covered previously. As a result, the following teams (or subsets thereof) typically exist one way or another:

- Astrodynamics and mission design (including models in, for example, AGI STK™)
- Systems engineering (roughly black-box-level design of all subsystems required to achieve the mission)
- Attitude and orbit determination and control[31]
- Guidance, navigation and control
- Accommodation (three-dimensional arrangement of all subsystems)
- Structure
- Thermal

---

[30] This book uses this definition, as well as the more contextual meaning of "operation of a product"

[31] They often also design other control loops, e.g. in mechanisms

- Radiation
- Radio Frequency (i.e. antennae)
- Optics
- Power generation (e.g. solar arrays)
- Energy storage (i.e. batteries)
- Electronics[32] hardware
- Software
  - Low-level software, i.e. firmware (close to hardware)
  - High-level software (for more advanced tasks)
- Harness
- Mechanisms
- Propulsion
- Ground Support Equipment
- Reliability
- Robotics
- Signal and data processing and analysis

Manufacturing, Assembly, Integration, and Tests has a significant fraction of its staff that is able to be assigned to any project on the production floor. Some areas do require very specific skills though, such as[33]:

---

[32] Called Avionics for the platform, for "Aviation electronics" which was extended to spacecrafts

[33] Others may benefit from specific expertise, but their

- Precision machine shops / workshops
- Metrology facilities (i.e. precise measurement of parts).

This often makes a case for subdividing MAIT into Manufacturing on the one hand, and AIT on the other.

Depending on the company's core business, some or even all of the MAIT may actually be subcontracted (still following this "make or buy?" ubiquitous question) – especially parts manufacturing or environmental testing. In fact, even in the design office, analyses are often subcontracted to subcontracting agencies. In this case, "sold" analysts are still required to use the company's IT infrastructure to prevent compromising confidential data.

Specialities

Teams regularly have to look into (figuratively-) microscopic details of their area of expertise to prove their design will work, and as a result every team generally identifies specialities that are worth

---

repeatable nature makes it optional as long as theoretical specialists from the design office supervise the process (e.g. environmental tests).

assigning sub-teams or at least individuals to. You can assume that every single element of the spacecraft will have been investigated in great detail at one point or another (more likely continuously) by a few specialists.

Fasteners, for example, spring to mind because the general public often refers to them as "just screws/bolts" whereas fasteners represent one of the critical parts of a spacecraft which undergo full theoretical and experimental characterisation, careful selection, thorough justification, and systematic control.

ROLES & RESPONSIBILITIES

What roles typically exist?

Who could I imagine to be?

We have now arrived at the base of any company: individuals. Akin to bees and ants, every individual has a place, their part to play: their role. Roles map duties and responsibilities to people (who, and I know how it sounds, should be interchangeable[34]). "Hi, I'm X, the new mechanical analyst on project Y" instantly tells you what technical jargon you can use, what they do or will do, and what they are accountable for. This role may be fixed and identical to the job title/position, or suitably defined for each project[35]. So much so that most if not all inter-teams or inter-company meetings start with a round of introductions. Either way, the definitions yet again vary from one company to another, but standards, interactions and convenience thankfully resulted in generally consistent roles with only nuances in

---

[34] After all, what happens to the project if someone leaves prematurely?

[35] More explicitly: roles can also be cumulated

responsibilities and naming.

You will generally find:

- Board of executives
  - Chief Executive Officer: the highest-ranking executive. Responsible for overall strategic management and decision-making in the company.
  - Chief Operations Officer: oversees and optimises the company's day-to-day operations
  - Chief Financial Officer: oversees and optimises the company's finances, past, present and future.
  - Chief Technical Officer: oversees the company's technical work and strategy
- Department director: oversees the operations of a specific department, setting goals, and managing team leaders
- Manager: generic title that can apply in any department or any team, referring to a position of direct or indirect responsibility for resources such as staff or budget. Typically plans, organises, coordinates, reports etc.

- Team leader: specific case of Managers. Leads a team of individuals sharing common skills or goals, assigning tasks, monitoring their progress, and facilitating collaboration. Most employees work under a team leader.
- Officer: refers to a position which has some degree of enforcement of rules by cross-departments authority, such as Quality Officer or Health and Safety Officer.
- Hired third parties:
  - Consultant/counsellor: leverages their experience to provide on-request advice, guidance or feedback
  - (Sub)contractor: delivers products, or services for a predetermined work package or period of time. May be integrated on-site.
- As for team roles, specifically in engineering (may apply elsewhere):
  - Program manager: leads and manages the planning, development, and execution of a large mission, or a series of missions, ensuring their successful implementation.
  - Product owner: responsible for the overall success of a specific product,

focussing on the purpose and definition of the product itself

o Project manager: responsible for the delivery of a product or service meeting the set requirements, within the lead time and budget allocated

o Responsible Engineer: combines the project manager role for a specific project (where achievable) or work package, with its technical leadership or even execution. Loosely speaking a compromise between the stereotypical (bad) project manager who may ask that you talk exclusively in months and dollars, and the stereotypical (bad) technical staff who may not care for either. Note the term "compromise", due to limited time.

o Product architect: writes and interprets requirements, designs architectures and development plans, and controls compliance to requirements

o Focal point: appointed individual that centralises all matters related to a specific strategic component or aspect of the design under the responsibility of a team, to build up a specialist status

o Specialist: not quite an expert, but somewhat halfway there

o Expert: recognized as a trusted technical authority[36] thanks to extensive knowledge and experience in a specific field

o Analyst: designs and runs analyses, usually on advanced analysis tools, to provide requested reports

o Design Engineer[37]: imagines, designs, and justifies solutions to requirements. Often able to make and test breadboards and prototypes.

o Manufacturing Engineer: turns drawings and other manufacturing requirements into actual parts (e.g. machinist)

---

[36] Some companies even use Technical Authority for this role when Expert is reserved e.g. for Senior positions

[37] In English, Engineer can designate either a design engineer or any other engineer who might not have the analytical background to do advanced design and justification. The latter, often referred to as Technicians, are just as valuable because they accumulate firsthand experience and intuitive understanding of the limits of designs.

- Maintenance engineer: maintains, diagnoses and, when possible, repairs support equipment
- Process engineer: imagines and designs plans and procedures
- Test engineer: Responsible for carrying out tests, and designing them if they are uncommon
- Operator: executes procedures, which may involve operating equipment. Can be a contextual role (e.g. Test engineers also operate).
- AIT Engineer: at least an Operator, often also a Test engineer, and sometimes a Process Engineer or even a Design Engineer (mainly for support equipment)

...

Military ranks may complement these roles even outside of military applications, e.g. for astronauts.

No doubt that many colleagues will not agree with the roles listed or their general description, but accuracy is irrelevant. What matters is that you:

- Realise many roles are typically involved in the Space industry

- Understand the mindset behind how and why roles are created

Always try to understand the roles and titles of those you will or may be interacting with – which can be as simple as asking them. As for yours, if your job title or project role is not accurate, you may even be free to define how you see yourself and get it approved[38] - as long as it clearly reflects what you actually do or can do. I see it as "finding your place" – to everyone their own.

Exercise: go back to §"Ecosystem" and try to imagine specific roles in which you would see yourself working, in each of the types of entities presented.

---

[38] Regardless of whether it is for a project role, or a job title, the matching description of tasks and responsibilities must be defined. The role name just an alias, a shorthand after all.

WORK DYNAMICS

How do missions and projects unfold?
How do they influence the day-to-day work?

Work in the Space industry is intertwined with missions as they come to life, evolve, and eventually conclude. With each mission, the nature of the work evolves, impacting everyone involved. Then, as a new mission emerges, the cycle begins anew, presenting a completely different set of products, challenges, and opportunities. This ever-changing landscape ensures that the work in the Space industry remains dynamic and distinct at any given point in time.

Mission phases and milestones

Missions and programs unfurl through distinct phases, akin to acts in a performance. Although standards may differ from one supervising agency to another, they generally align well to facilitate international collaboration. The European Space Agency's definition of those phases and expected key milestones are:

- Phase 0, Mission analysis and needs identification: the mission itself and its concept of operation are defined and translated into practical needs.
  - Road trip analogy: "Where would I suggest to go? Why?"
  - Key milestone: Mission Design Review (MDR)
- Phase A, Feasibility: the high-level aspects of the mission, including performance, budget, and development time, are evaluated to determine if it is feasible.
  - Road trip analogy: "Is everyone available that month? Do they like what I envision?"
  - Key milestone: Preliminary Requirements Review (PRR)
- Phase B, Preliminary definition: a preliminary design of the technical solution to the mission needs is defined and justified[39].
  - Road trip analogy: "How are we going? What week? Can we afford this?"
  - Key milestones:
    - (final) System Requirements Review (SRR)

---

[39] ~ demonstrated that it should work

- Preliminary Design Review (PDR)
- Phase C, Detailed definition: the final, complete, design of the technical solution to the mission needs is defined and justified. Key milestone:
  - Road trip analogy: iterating on and detailing the final itinerary, budget, and dates
  - Critical Design Review (CDR)
- Phase D, Qualification[40] and production: units are produced and tested such that the technical solution is delivered with high confidence that it actually meets the mission needs.
  - Road trip analogy:
    - Qualification: none or by "delta-qualification", which is very loosely "I know this works because I've done it in the past, or something similar and I just need to check that [...]"
    - Production: Book everything, except critical items that were decided to be booked long ago

---

[40] C.f. §"Models" below

to avoid large price increases. Pack.

- o Key milestones:
    - Qualification (results) Review (QR)
    - (flight model) Acceptance[41] Review (AR)
    - Operations Readiness Review (ORR)
- Phase E, Utilisation: the mission is launched and enters the operational phase, where it fulfils its purpose.
    - o Road trip analogy: meet and go.
    - o Key milestones:
        - Flight Readiness Review (FRR)
        - Launch Readiness Review (LRR)
        - Commissioning[42] Results Review (CRR)
        - End of Life Review (ELR)
- Phase F, Disposal: at the end of its operational life, the mission is decommissioned or disposed of in a controlled manner.

---

[41] C.f. §"Models" below
[42] First steps after injection in orbit, usually until entering actual operation

o Road trip analogy: saying goodbye, sharing photos, etc.
o Key milestone:
  ▪ Mission Completion Review

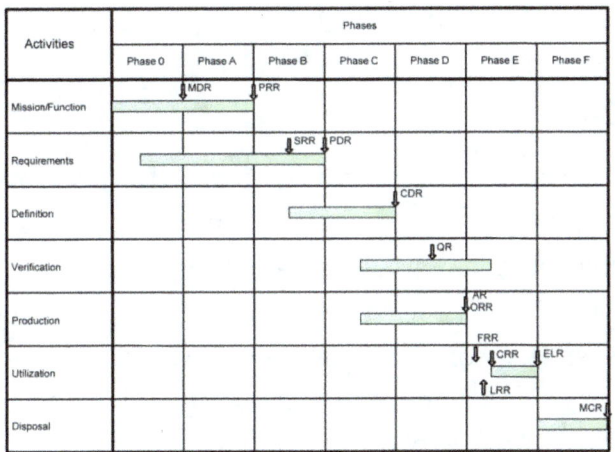

Figure 33: Typical mission/program life cycle (from ESA's ECSS-M-ST-10C)

The table below presents a crude rule of thumb of when major actors tend to be involved the most throughout the mission phases.

	Phases						
	0	A	B	C	D	E	F
Mission stakeholders	▮	▮				▮	
Space agencies	▮	▮				▮	
Space & Ground systems manufacturers			▮	▮	▮		
Launch site & spacecraft operators						▮	▮

▮ = most involvement

Figure 34: Typical involvement of major actors throughout a mission

Note that stakeholders can cover a wide range of companies and individuals, depending on the mission (c.f. §Ecosystem).

Models

Although our understanding of the laws of the Universe has progressed immensely, it is still impossible today to design a new Space product and get it right first try. The engineering process necessitates some trial and error when it comes to building, and since building is long and expensive, a limited number of models of increasing fidelity is planned where it matters the most.

Models vary from one project to another since they are specifically defined as part of a de-risking strategy. Nonetheless, one can generally find:

- Breadboard models: do not look anything like the product considered, but are very quick to make and yet still allow to mitigate some risks. For example:
    - Electronics on an actual breadboard / perforated boards

- o Plumbing arrangement using industrial components
- o 3D printed model (e.g. to test accommodation or assembly)

  …

- Engineering models: de-scope some of the product, focussing on only certain critical aspects. In other words, "fakes" a certain aspect of the design. For example:
  - o Structurally representative (e.g. "dummy mass")
  - o Thermally representative
  - o Electrically representative (e.g. "Flat Sat" where all components are laid flat, c.f. Figure 29)
  - o Optically representative (e.g. optics testbench)
  - o Downscaled
  - o No electronics

  Engineering models can cumulate multiple of these aspects, e.g. Structurally & thermally representative.

- Qualification models: identical to flight models, but do not fly. They undergo a thorough, expensive, and long test campaign, and since their design is identical to that of the flight models, all results are assumed to be applicable to all flight models as long as the flight models built are actually sufficiently similar to the qualification model (c.f. below)

- Flight models: are deployed in Space after the more limited acceptance tests (in labour, and levels), which are designed to prove the similarity of the built models to the qualification model. This includes workmanship, meaning whether the model has been correctly built.

Note that not all projects result in a flight model. Some projects may stop at engineering or qualification models, until further funding is secured.

Technology maturation

Spacecraft manufacturers only use components and technology which have extensive flight heritage – except when they cannot. This occurs primarily in instruments where new technology actually enables new missions and programs. Moreover, new technology can also be an asset over competitors. Regardless, there is significant research and development effort to mature technology until engineers can use them for flight.

Maturation of technology is measured with the Technology Readiness Level, shorthanded TRL. NASA and ESA both have specific definitions for the technology readiness levels, but here is an approximation[43]:

- TRL 1: idea or a basic concept that has not been tested or proven yet
- TRL 2: the concept has been studied further to determine if it is technically feasible and how it may be used
- TRL 3: the basic functionality of the technology is assessed using prototypes that are far from the final product and in a laboratory environment
- TRL 4: the performance of the technology is as expected, as tested on improved models in the same laboratory environment
- TRL 5: the performance of the technology is as expected, but this time as tested in a relevant environment[44]
- TRL 6: a flight demonstrator has successfully been tested in a relevant environment

---

[43] Levels have been modified slightly for simplification
[44] "Relevant" means here an environment which is sufficiently close to the application's to be able to say "this model would perform similarly well up there" when the tests pass

- TRL 7: The technology has been demonstrated in Space, but has not been used for actual operations
- TRL 8: The final product using this technology has successfully passed flight qualification
- TRL 9: The final product has successfully been used in operations in Space

TRL ratings under 4 are rarely considered by spacecraft manufacturers to be included in a new mission, if only for technology demonstration missions. Regardless, it is common for technology to originate in academia and advance through industry-driven maturation. The figure below illustrates the progression.

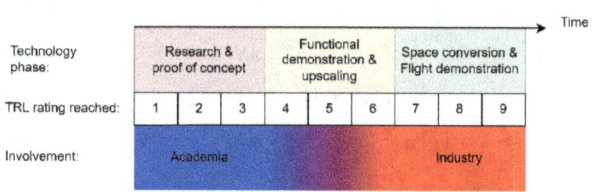

Figure 35: Technology development progress

The TRL is not to be mistaken with the qualification status of a spacecraft component. The qualification status follows internal standards aiming at describing, without numbers, how much additional testing is required for the considered component to be allowed to be integrated into the targeted spacecraft, based on previous tests and missions. It is therefore assessed ad-hoc. For example, "Category X: delta-qualification required" (specific tests to be defined).

Development cycle

Development projects commonly follow a structured approach known as the "V cycle." This approach ensures that requirements are consistently addressed during development by aligning different levels of design definition with corresponding compliance checks. In other words, we define what we want, and always check that we actually had what we wanted when we get it:

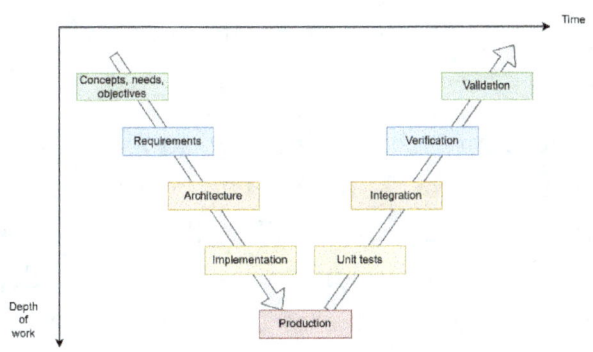

Figure 36: V-cycle development approach

- Concepts, needs, and objectives of the product are defined ahead of requirements and are all validated at the end of the development before delivery
- Requirements are flowed down from the concepts, needs and objectives, ahead of defining the product architecture, and are all verified before final validation.
- The architecture of the product is defined based on the requirements, ahead of its implementation, and it is demonstrated by running integration tests.
- The implementation of the architecture is designed before production, and is demonstrated at unit level (lowest level of testing possible) before integration of all units

The V-cycle approach is generally adopted by all teams, meaning that there are even layers of entire cycles (i.e. nested cycles like Russian dolls). Moreover, design iterations and various models mean that a single team goes somewhat back and forth in the V-cycle and still completes it several times per project.

Every individual, regardless of their role, feels the various stages of the V-cycle they are in, in the nature and the depth of the tasks at hand. Reviews close out each stage much like the conclusion of a theatrical performance.

Key documentation

The Space sector is among the most documented on the planet, with the military[45], the medical, and the nuclear industries. Some might see it as an obstacle, but aside from being a natural part of the job (c.f. §"Corporate rules"), it is actually a strength: everything we ever need is, or will be, written somewhere logical. It is what allows exceptionally large consortiums spread across multiple continents, to deliver a single product so seamlessly.

There is no point in trying to list a large number of documents. Instead, here is a list of key documents which stand out in my opinion as being extremely important.

- Requirements: list of needs[46] defining what a deliverable has to do or be. Requirements flow down like a waterfall from a source (e.g. the

---

[45] Imagine when both are combined
[46] Requirements are specific, measurable, agreed, realistic and timely (SMART).

mission requirements) – though they can be negotiated back up if necessary. They exist at every interface, so they can also describe an activity (e.g. assembly, tests) or a service.

- Interface Control Documents: play another vital role in ensuring smooth collaboration between teams by providing a black-box equivalent to a team's deliverable before it is completed. Basically what the product is from the outside. This allows teams to work independently, confident that their deliverables will seamlessly integrate. For example, mechanical ICD (inc. drawing(s)), electrical ICD (inc. pin assignments)...

- Specifications/datasheets: as their name suggests, sheets listing (hopefully) a lot of data about the product. Particularly useful for components selection and inclusion in the design. Basically tells what the product can do.

- Compliance documents: Set of documents listing which requirements were met, not met, partially met, and why.

- Proposal: usually significant to large undertaking describing a technical solution, its cost, timeline, and development framework, with the aim of obtaining the resources to make it happen. Often made in response to a call (e.g. Invitations To Tender or Requests For Proposals).

- Review slideshows: Slides presented during a review. Since reviews are key milestones where designs are presented in an organised and summarised way, so review slideshows are particularly convenient to get up to speed efficiently.
- Design descriptions: describe qualitatively the design with the aim of understanding its working principle and thought process. Sometimes, the review slideshows can cover this part.
- Design justification documents: detailed demonstration that every aspect of the design does what it should do, generally including key budgets for design drivers (e.g. mass, power, error...)
- Schematics and diagrams: visual representations that show how different parts or components of a system are connected
- Plans: outline the intended strategy or approach, to achieve specific goals (e.g.

development, safety...). May call existing procedures.

- Procedures: step-by-step instructions for carrying out specific tasks or activities (e.g. assembly, tests).
- Reports: documents that present results and conclusions of carried out activities (e.g. analysis, tests, non-conformance). Include assumptions and methodology if a procedure was not followed.
- Declared lists: provide information about regulated or controlled aspects of the product such as materials, components, and processes. Particularly useful to QA to ensure no part of the design is forbidden or loosely controlled.
- Standards: set of guidelines and requirements made internally or by a third party, defining a specific level of quality or performance. Standards can be optional, or made mandatory by law or company policy. In all cases the requirements capture which have to be followed.
- Quote: provides a price tag and lead time for a special request of products or services. Often

one of the most anticipated documents, albeit one of the simplest (to read).

- Contract: legally binding agreement that outlines the terms, conditions, and responsibilities related to the project
- Statement of work: explains what needs to be done for a project, including the goals, tasks, and expected results. Basically the technical part of a contract.

Key documents and files are always *configured* in a content management system, i.e. they are uploaded onto a secure server where it is made impossible to accidentally modify or delete them, mix up versions, or release an unreviewed or unapproved version. Accesses are granted on a need-to-know basis, but available easily from anywhere in the company network based on their unique identification number and name.

ROLE PLAY

What would it be like if I worked on a typical project?

The following example is purely fictitious, designed for the sole purpose of this exercise. Any resemblance with a real project is coincidental.

Imagine that a systems/applications architect working for a satellite manufacturer has invented a concept for a novel instrument that would be an asset for the nation's defence, yet which also has great potential in civilian applications (e.g. environmental forecast). Potential customers from both applications are identified, approached, and they show interest, but the technology has to be matured before it is considered for an in-orbit demonstration mission. Specifically, a challenging single-axis pointing mechanism at the heart of the concept – for which there is no precedent, no heritage. A proposal is written to apply for partial funding of the development of the mechanism, up to a qualification model (if possible – or an Engineering-Qualification Model, at least, which is the next best thing), via a grant program offered by the nation's army department. The proposal is selected, and the contract signed, such that both parties pay for half of the total cost - after all, this mechanism would later become part of the company's products range and eventually bring business.

Meanwhile, in the mechanisms team: at a weekly team meeting, the team leader announces the proposal is coming their way, along with some other projects, and opens a discussion to dispatch the new arrivals in the team's work plans. Some members may be more available due to projects ramping down, but interest and matching skillsets are other constraints that need to be taken into account. Indeed, ideally, the person in charge of the proposal will also drive the project if/when it is approved/selected.

Today, that person is you.

You have derived requirements for the mechanism from the instrument's requirements, thought about all the major risks of the project based on the team's mechanisms heritage and common experience, devised a sound development strategy to guarantee the delivery of the requested product in spite of those risks, translated it in hours and direct costs, and got it all approved. First by your team, then by your internal customer (the instrument's architect now doubling as a product owner). It took great teamwork and a few iterations, but the proposal is finally sent and temporarily out of the picture while it is being processed. This gives you a boost as you turn to another project, since it is usual to juggle 2 to 4 projects at a time like pans on a hob to make sure no time is ever wasted. If the proposal is eventually greenlit, the development plan so carefully designed and described in the proposal is going to be put to the test.

Literally. The total cost of the project stated in the proposal, reflecting your entire plan, ought to be accurate[47]: a fixed wallet is created for all expenses and all hours booked by everyone expected to be working on the project, until delivery. It may not be your money, but this is now your wallet. For 2 years worth of work for everyone involved, in this case. Sub-wallets are also created with the distribution you had in the proposal's cost breakdown, and you will continuously track and report their usage (matching the work packages you planned) to ensure the project will not run out before the end. For reference, you may use the proposal, which should also break down the cost by milestones and project phases.

---

[47] ideally conservative, but it decreases the chances of obtaining the funds

The mechanism is a seemingly simplistic one-axis mirror pointing mechanism, i.e. basically a rotating shaft, balanced[48], driven by a motor to follow specific profiles/trajectories based on an absolute angular position sensor. It is intended to be part of an instrument mounted on a geostationary satellite to get a global, static, view of the Earth. The mechanism's main technical challenge lies in the fact that it is very accurate, very fast, and long-life. A challenging combination because a) fast movements excite resonances which create oscillations, and b) at the speeds targeted, the typical service life of 15 years could easily mean broken pivots from fatigue. Only compliant[49] pivots would work, which means that all other components cannot be off-the-shelf components with integrated bearings, as that would introduce too many constraints[50]. And yet, said other components require very tight and stable gaps between stator and rotor. So not only everything must be custom-made because no one makes what

---

[48] Meaning the shaft is designed such that it does not generate vibrations at it rotates (contrary to the shaft of a smartphone's motors)

[49] Compliant roughly means "which deforms". A compliant pivot is a single part, designed to deform such that either side rotates around a fixed common axis. Compliant parts may be designed to have "infinite life".

[50] This is called "hyperstatic": a four-legged chair is hyperstatic, if you nail each leg to an uneven floor, the chair will deform and build up internal stress. The stiffer the parts, the worse it gets.

you need (including your company), but even once made, integrating everything will also be a challenge. Of course, you already know all of that at this point, because you wrote it with your collaborators in the proposal.

It is now time to kick off the project with the team identified in the proposal: in the early stage, you will mainly be working with:

- Colleagues from the mechanisms team, who are focal points for key aspects of the design:
    o Motor - to drive the design of the motor component.
    o Position sensors - to drive the design of the position sensor component.
    o Compliant pivots - to assist in accommodating the pivots (we will assume here they already exist)
    o Mechanisms analyst - to estimate the behaviour of the mechanism to high-speed, vibration, and temperature effects
- Mechanical designer - to define and 3D-model the mechanical design of the mechanism.
- Thermal analyst - to make a thermal model of the mechanism and of its environment,

calculate all temperatures and select thermal control hardware

- Mechanical analyst - to make a structural and thermoelastic model of the mechanism to calculate stresses, alignments, or mechanical relationships
- Control engineer - to design a controller for the mechanism, i.e. the "pilot" if the rest were indicators and controls.

... Backed up by a Senior expert from the mechanisms team. You will however seek assistance from many other disciplines throughout the project, for example optics, because the mirror is taken into account even if it is not part of the defined scope of the mechanism.

The next step is to compare various concepts and options that may have been mentioned in the proposal - namely here, the arrangement of the components along the axis, or the technology of the components. All combinations are listed and scored, as a team, and as per trade-off criteria specifically selected for this project. A few options are agreed to be worth exploring. These few options are further explored, with crude 3D models, dynamics analyses, and by deriving driving requirements for components for example. You will soon realise which requirements are major obstacles, and turn to the architect or product owner to see if and how they could be relaxed. At some point in this phase though, a Requirements Review will mark the end of these early negotiations to allow things to settle.

At the end of the trade-off, a concept comes out on top - about time because looking at your overall timeline, the clock is ticking fast: the components are long-lead items of which the development should start now to stay on track, based on first rough quotes. Preliminary requirements for these components are therefore made as conservatively as possible, to be able to start calling for subcontractors while the design has not converged yet.

The baseline of the mechanism, i.e. what it looks like and is required/expected to do, evolves from one week to another - including back and forth - until the general architecture has converged. Yet, subcontractors are by now selected for all components and kept in the loop of requirements changes, including for the manufacturing and assembly that is here also subcontracted. This is a sign that every contributor is confident they can meet their own requirements. It is therefore time to hold a Preliminary Design Review, to make sure all stakeholders are satisfied about the direction the team is going before freezing the architecture. It is somewhat late compared to your initial plan, but the timeline evolved with the project and all stakeholders are in line with your current plan. The team shifts their effort to documentation, hoping there are no hidden major issues, or that if there are, that the review board picks up on them. For this reason the review is thorough, concise, and sent as a pack of documents ahead of the meeting such that the reviewers may have time to prepare the meeting. Some recommendations are made the day of the meeting, some others after the meeting, but the review is finally closed out and greenlit for the next phase: detailed design.

Whereas requirements definition and preliminary design phases are volatile in that any aspect of the project may be completely different the next week, in the detailed design phase, all interfaces and requirements are well-defined, describing mature black boxes. They vary much less frequently, and the underlying design is therefore slowly refined more and more with time, with less back-and-forth. Details such as harness are added. Placeholders and envelopes make way to calculated and optimised values. Analyses (performance, thermal, mechanical, radiation) are no longer partial nor targeted on major risks, but instead are covering all relevant requirements. Margins are tuned down as confidence in unknowns increases. Support equipment, dummy parts[51], and features are designed[52] as manufacturing, assembly, transport and testing are planned out. At the end of this phase, you have thought about and defined everything you (as a team) could (or deemed worth doing), because significant expenses will be committed once the plan is set in motion: in this project, most of the risk is taken here because there were no earlier models. Indeed, additional Engineering Models or breadboards were replaced with lengthy analyses given how interdependent and costly most of the mechanism is. The milestone where stakeholders decide whether the project is ready to be materialised is the Critical Design Review. This time, the documentation and review meeting content is significantly more involved than for the Preliminary Design Review, but thankfully the justification of the design should have been done more in "real time" in this phase than in the previous phase. After the go-ahead, the entire design is translated into the documents, drawings and files the manufacturers need to make the mechanism. Special care is given to

---

[51] For example here, a thermally-representative mirror, or a mechanically-representative mirror
[52] Note that custom EGSEs may start being developed in the preliminary design phase, depending on their complexity

tolerances on drawings, so that the assembly and the performance of the mechanism are achieved as long as all parts produced are within the set tolerances.

After the Manufacturing Readiness Review that greenlights the order, it typically takes months to make parts this complex - which leaves enough time to detail the procedures that will have to be followed for assembly and testing. Since assembly is here subcontracted, the focus would be on testing: regardless of whether the environmental tests are also subcontracted or not, all functional and performance tests (of which some are executed during environmental testing) are under your responsibility. You might nonetheless have more time for other projects while the mechanism is being made - including planning your time after the project's end.

Once the mechanism is on the table though, months of planning unfold in a few weeks. Since no plan and procedure is perfect, this is a particularly intense period - but a rewarding one nonetheless: if you were the mechanism, you would have to climb into a washing machine, go through a wash cycle, get out, sit in a rolling chair that alternates between a biting cold room and a scorching oven-grill, while performing clockwork repairs on your lap knowing full-well you have to beat a challenging timer or you're fired. Oh, and then run a marathon when you're done. Yet, your little one delivers. Tirelessly, it waltzes through the challenges and checks all the boxes. You might have a few disappointments, but it is also the opportunity for an interesting puzzle full of valuable lessons. Either way, after some time unpacking and analysing the large amount of resulting data, you can finally say with confidence what your mechanism can do in orbit and, if it falls short of the requirements in certain areas, suggest solutions. This is the Tests Review Board. For an actual qualification model, depending on your risk-taking strategy, this could be greenlighting (or delaying) the production of flight models - but it simply means here that the project can take a short well-deserved rest until it is tapped on the shoulder for an actual mission.

Of course, this is a textbook, ideal, and extremely simplified example. Real life is much more complex and challenging, which is actually a good thing: projects may resist you like never before, but it is all the more rewarding when they finally crack open and become cutting edge technology. Moreover, delving into them from various angles ensures that with each new project comes an ever-growing depth of acquired knowledge.

CORPORATE RULES

What would I be expected or required to think about at work?

The following is a selection of important rules which are expected or even required to be followed, yet can easily be overlooked or forgotten. Note this is not necessarily specific to the Space industry – it is provided for completeness.

1. **Control disclosure**. Basically, ask yourself: "Am I allowed to say/give this?".
    a. You have to ask yourself whether you are giving confidential information or even patented intellectual property without a suitable non-disclosure agreement in place (which legally forbids any leak of information by whoever you are talking to).
    b. Likewise, within your company, you may only be allowed to talk about certain subjects / projects with staff that have a sufficient "clearance" level, within the walls of special rooms with

restricted access. Files on such topics would also have special treatment.

2. **Control exports**. You have to ask yourself whether you are exporting controlled goods or information[53] without a suitable license in place (which can be as simple as working from abroad on sensitive information)

3. **Prevent industrial espionage**. You have to ask yourself whether you may be compromising your work station (for example by working in a public space without screen filters)[54]

4. **Only take authorised risks**. Ask the right people in the most efficient way possible either for confirmation, or for instructions when in doubt. If risks are taken without following the process in place, an investigation will be open – and losing a job, getting a fine, or even jail time could result from it for you or someone else higher up. See 6.a.

5. **Prevent conflicts of interest.** You have to ask yourself whether you are in a conflict of interest, i.e. if you are susceptible to be biased, to avoid corruption. For example:

---

[53] Controlled by governments, e.g. technology that can be used in weapons

[54] Another example as a funny anecdote: hackers would previously leave re-packaged corrupted flashdrives outside Space companies' entrance turnstiles. As a result, personal flashdrives are often forbidden.

a. When selecting providers you might know personally
b. When being offered gifts

6. **Document**. You have to ask yourself what documents are necessary to write in your day-to-day job to:
   a. Justify the absence of negligence in case of an issue;
   b. Enable someone else to help, or take over in case you leave;
   c. Get your work reviewed (which should be always)

7. **Think as a company**. You work for a company. It is not about the time you save yourself, but the time and cost you save the company. Find what the company (if it was a person) would want you to do and not do. This starts with understanding your chain of command and overall company organisation, but it also involves taking into account what your colleagues' workload and scope of work are.

8. **No negative emotions**. You are required to control your emotions. Positive emotions are encouraged, but negative emotions are expected to not exist in the workplace. Professionals communicate factually and cordially. Talk to your supervisor or to human resources if this is an issue.

9. **Facts. No bias.** The workplace requires you to take decisions based on (potentially changing) facts, without bias. In particular, you may like working with some people more than others or prefer to avoid some, but this cannot influence your actions. Staff members have only one label: role(s) or position. Talk to your supervisor or to human resources if this is an issue.

10. **You are an ambassador of your company.** Remember that some of your actions may be associated with your employer – even outside of the workplace –, and the image of a company strongly influences its health. For example, if you want to give your opinion on a sensitive matter in public, add a disclaimer saying your company does not necessarily share it.

11. **Communicate.** Always ask yourself whether a recent update should be forwarded to someone else, in particular a delay in a deadline, or a cost overrun – however likely it might be. You might not know the extent of their impact in the big picture.

12. **Prioritise.** You are expected to prioritise your work: delays occur regularly, but they are accepted only if it could not have happened any other way. If you are unsure, ask you supervisor to confirm your priorities.

13. **Keep track of hours.** Projects often run on a fixed budget including labour costs, meaning hours spent on each project are booked to that specific project. Improperly filling in your time sheets is essentially like moving money between projects – which in this case, would be forbidden, and eventually detected by project managers. Though there is reasonable tolerance.[55]

14. **Never hide anything nor lie.** Come forward as soon as possible instead. If you are afraid of consequences for reporting an issue, things are seriously wrong in your department and should be fixed as soon as possible for the sake of the company.

---

[55] For companies that do not, it is still valuable to roughly track time because it improves your ability (or someone else's) to plan future projects.

ADVICE

Any tips or tricks?

Some tips I like to give regularly. None of this is required, but I recommend those all the same.

1. **"There is no free lunch."** If something seems too great to be true, chances are it is. Everything has a cost and weaknesses. Only their relevance and relative impact can be changed by suitable matching.

2. **Plan your development**: always list your risks first, and start from there. What scares you the most? How would you be reassured as early as possible? Not only at the start: regularly stop, look ahead, and plan where you are going. Your journey will be a lot more efficient if you do not look at your feet the entire way.

3. **Divide and conquer.** A good modular architecture makes the resulting work packages easier to complete and to track, and facilitates collaboration.

4. **Prepare meetings**. All meetings - think about how much it costs hourly for that many people. If you organise them, make them efficient. If you attend: do not improvise if you were able to prepare (unless you are both concise and good at improvising).

5. **Understand the filing system**, even for non-configured items. Messy project drives are a nightmare to use, and there is almost always a standard or guideline to follow (or someone to ask).

6. **Be curious about what you do not know**. Not only in your field, but also your colleagues'. It does not have to take long, it is the accumulated effect that makes a difference.

7. **You are all in the same boat**, and that boat shall not sink. If you think something does not make sense, you are allowed to second guess or ask questions, as long as you do not waste the other person's time by not having done your research before. No one is always right after all.

8. **Any phenomenon you might consider negative is positive in a different context** – The laws of the Universe are never against us, we just have to use them right.

9. **Be precise, if possible with less words**. It should be impossible for anyone to wrongly interpret what you say or write. Shorter is still always better than longer, so good structure and pictures/diagrams should be preferred over equivalent words, but use as many words as it takes to stay precise. An ideal goal, but one from which we cannot afford to deviate. E.g. this book is not precise.

10. **Challenge requirements**. You might be wasting efforts trying to comply with a requirement which is obsolete or could be relaxed. Requirements are a large undertaking, they cannot reflect the best overall compromise without feedback.

AFTERWORD

I hope this book has answered the questions you had when you first opened it or, at least, that you have learned something from it. I will be keeping my ear to the ground for ways to improve it with the goal that it eventually does for everyone.

If you now have more questions than you initially had, this is actually excellent: questions are fuel, who knows where they will lead you?

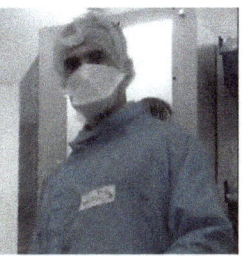

Grégoire Marchetaux is a space mechatronics and robotics architect who has been working in the space industry since 2013.

He has experienced working for:

- A research institute doubling as a space instrument manufacturer, in Harwell-Oxford's space centre in the UK (RAL Space)
- A Space industry leader, in France (Airbus)
- *(present)* A startup building a simpler and greener launcher for Canada (Reaction Dynamics)

As of 2023, Grégoire has worked:

- In the Robotics R&D, AOCS, Mechanisms, Avionics, and EGSE teams, mostly as Responsible Engineer, Product architect, and Specialist.
- Throughout all stages of development from research and proposal to delivery
- On 10 Space missions/programs of which 7 are already known to have flown.

Back cover background picture credits:

Lockheed Martin, 2021